THE LIBRARY OF WHY?™

Why Do the Oceans Have Tides?

Marian B. Jacobs, Ph.D.

The Rosen Publishing Group's
PowerKids Press™
New York

For my grandsons, Carlos and Gianni.

Published in 1999 by The Rosen Publishing Group, Inc.
29 East 21st Street, New York, NY 10010

First Edition

Book Design: Danielle Primiceri

Photo Credits: Cover © Stephen Simpson/FPG International; p. 4 Dick Dickinson/International Stock; p. 8 © Steve Hix/FPG International; p. 12 © Bill Terry/Viesti Associates, Inc.; p. 15 © Telegraph Colour Library/FPG International; p. 16 © Chad Ehlers/International Stock; p. 19 © Ken Reid/FPG International; p. 20 © John Michael/International Stock; p. 22 © 1996 PhotoDisc, Inc.

Jacobs, Marian B.
      Why does the ocean have tides? / by Marian B. Jacobs.
          p.   cm.— (The library of why)
      Includes index.
      Summary: Discusses the nature, causes, effects, and uses of tides.
      ISBN 0-8239-5272-X
      1. Tides—Juvenile literature. [1. Tides.] I. Title. II. Series: Jacobs, Marian B. Library of why.
      GC302.J33 1998
      551.47'08—dc21
                                                                                                    98-9261
                                                                                                       CIP
                                                                                                        AC

Manufactured in the United States of America

Contents

Why Are the Oceans Important?

Water covers almost ¾ of Earth's surface. Earth is the only planet in the solar system that has saltwater oceans and seas. Because we live on land, we may not realize how important our oceans are.

Together, the oceans and the **atmosphere** (AT-mus-feer) make our weather. They supply most of the water to the **water cycle** (WA-ter SY-kul) that makes clouds and rain. The oceans have lots of food, such as fish, clams, lobsters, and plants. Much of the food for people around the world comes from our oceans.

◀ *Not only does the ocean affect the weather and give us food, it gives us a place to cool off during the hot summer!*

Where Are the Oceans and the Seas?

The oceans are very large bodies of water that surround the **continents** (KON-tih-nents). Try to find them on a map of the world.

① Which ocean is the largest?

② Which ocean is shaped like an "S"?

③ Which ocean is south of India?

④ Which ocean is at the North Pole?

⑤ Seas and gulfs are like oceans, but they are smaller and are usually surrounded by land. Can you find a gulf and a sea that lie between North and South America?

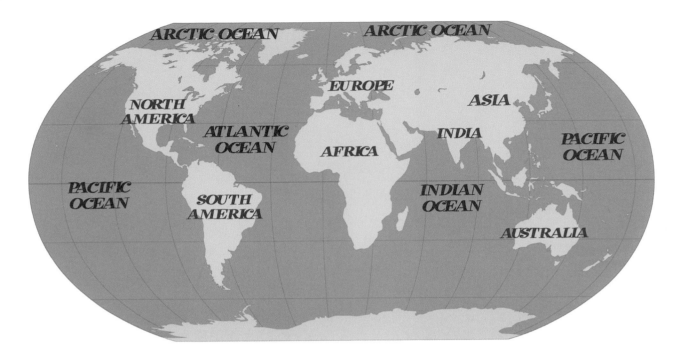

ARCTIC OCEAN

ARCTIC OCEAN

EUROPE

ASIA

NORTH
AMERICA

ATLANTIC
OCEAN

AFRICA

INDIA

PACIFIC
OCEAN

PACIFIC
OCEAN

SOUTH
AMERICA

INDIAN
OCEAN

AUSTRALIA

GULF OF
MEXICO

CARIBBEAN
SEA

What Are Tides?

Tides are the constant rise and fall of the sea level every day. During the day waves come in and slowly reach higher up on the beach. Sometimes you have to move your beach blanket to keep it from getting wet. The highest place the waves reach is called high tide.

When the waves slowly pull back, they uncover parts of the beach that were underwater during high tide. The lowest place on the beach where waves break is called low tide. This happens all over the world.

◀ The waves that break on the beach are not caused by the tides. They are caused by wind pushing on the surface of the water.

What Causes Tides?

Tides are caused by the pull of **gravity** (GRA-vuh-tee) from the Moon and the Sun on ocean water. The pull of the Moon is greater because it is closer to Earth than the Sun. The Moon pulls on the ocean water that is closest to it. This creates a bulge in the surface of the water, called a **tidal bulge** (TY-dul BULJ). The water on the opposite side of Earth also forms a tidal bulge. These bulges are what make high tide. Between the tidal bulges are lower levels of water which make low tide. This happens **gradually** (GRA-joo-lee). Most shores have two tides each day, some have only one.

The tidal bulges move slowly around Earth as the Moon does. Halfway between each high tide is low tide. ▶

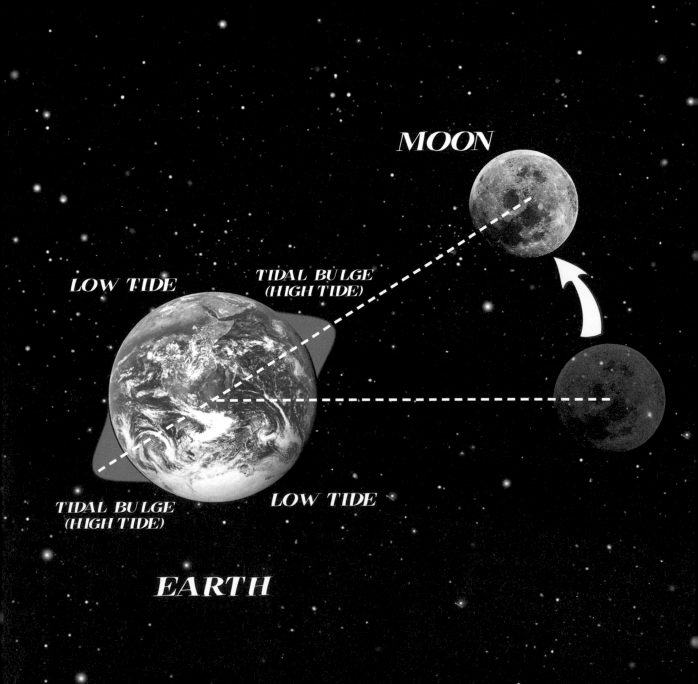

Are Tides the Same Everywhere?

Tidal **rhythms** (RIH-thumz) and their heights are different from ocean to ocean. Shores around the Atlantic Ocean have two high tides of about equal height and two low tides of about equal height each day. Shores on the Gulf of Mexico have only one tide each day.

When the tide rushes into a narrow bay where the water cannot spread out, the difference in height between high tide and low tide can be very large. This happens in the Bay of Fundy in Nova Scotia, Canada, where the highest tides in the world occur.

◀ *This is low tide in the Bay of Fundy. At high tide the water reaches all the way to the tops of those rocks!*

Do Rivers Have Tides?

Many rivers that flow into the ocean have tides. At the mouth of a river, water from the river mixes with water from the ocean and takes on the ocean's two tides every day.

Sometimes, when the tide is at its highest, a large wave from the ocean flows up a river. This is called a tidal **bore** (BOHR). Tidal bores only happen in some rivers. They can cause flooding along the river bank as they travel upstream. Tidal bores don't happen very often, but they are amazing to watch.

One of the largest tidal bores in the world moves up the Amazon River in South America. When the bore occurs, it can reach as high as 25 feet!

What Is the Tidal Zone?

The **tidal zone** (TY-dul ZOHN) is the part of the shore that is covered and uncovered by water because of the tides.

The rise and fall of ocean water has a great effect on the plants and animals that live along the shore. They must be able to live while covered with ocean water during long hours every day. Then they must be able to survive being out in the air, hot sun, and wind when the tide goes out. They also have to survive being hunted by other animals, such as birds.

Marine animals such as periwinkle snails, barnacles, mussels, clams, and starfish live in the tidal zone. Green algae and seaweed grow on the rocks there.

How Do Animals Survive in the Tidal Zone?

Animals have special features to help them live in the tidal zone. Some animals have thick, hard shells. Barnacles stick themselves to rocks and docks. Their grip is so strong they can stay attached even in rough waves. Mussels make a strong cord to attach themselves to rocks and each other so they don't get smashed by pounding waves.

Some animals burrow into the sand for protection. Clams have a little foot-like part that helps them burrow. Crabs use their claws and legs to dig into the sand and cover themselves.

Starfish have hundreds of tiny feet with suckers on them that help them cling to rocks in the tidal zone. ▶

Can We Use Energy from Tides?

There is lots of **energy** (EH-ner-jee) in the movement of the tides. People around the world use this energy to make **electricity** (ih-lek-TRIH-sih-tee). More than twenty years ago, the French built a dam across the Rance River in France for a new kind of power station. There, 24 **turbines** (TER-bynz) turn as water flows past with the change of tides. Electricity is created as the turbines turn. Other tidal power stations have been built in China and Russia.

Tides are an endless source of energy. They can make electricity with little pollution.

◀ *The Hoover Dam in Arizona and Nevada uses energy from water flowing in the Colorado River to create electricity.*

How Do Tides Affect People?

Tides affect people in many ways. Fishermen and fisherwomen watch the tides to figure out the best times to fish. Some wait for low tide. Then they can get the clams that are buried. Large ships wait for high tide to enter or exit shallow ports. Dangerous flooding can happen if high tide occurs during windy, stormy weather. The worst damage to beaches occurs during hurricanes at high tide.

The change in tides is a part of nature that scientists are still studying. The more we learn, the better we can understand our planet Earth.

Glossary

atmosphere (AT-mus-feer) The layer of air that surrounds Earth.

bore (BOHR) A large wave that travels up a river.

continent (KON-tih-nent) A main land mass on Earth.

electricity (ih-lek-TRIH-sih-tee) A form of energy that can produce light, heat, or motion.

energy (EH-ner-jee) The ability to do work.

gradual (GRA-joo-ul) To happen in slow stages instead of all at once.

gravity (GRA-vuh-tee) The force that comes from a body that pulls things toward it, such as the Moon's pull on Earth's sea level.

rhythm (RIH-thum) The movement of something with a regular rise and fall that repeats.

tidal bulge (TY-dul BULJ) The upward swell of sea level resulting from the pull of the Moon and the Sun.

tidal zone (TY-dul ZOHN) The area along a shoreline that is covered and uncovered by sea water in response to the tides.

turbine (TER-byn) A motor that turns by a flow of water.

water cycle (WA-ter SY-kul) When water evaporates from the oceans into the sky, where it forms clouds and falls back to Earth as rain.

Web Sites:

You can learn more about the tides at this Web site:
http://www2.gasou.edu/geol/8.3FTB.html

23

Index